Washington Ayer

Transitory Mania, with its Medico-Legal Bearing

A Prize Essay, Read before the Medical Society of the State of California

Washington Ayer

Transitory Mania, with its Medico-Legal Bearing
A Prize Essay, Read before the Medical Society of the State of California

ISBN/EAN: 9783337187583

Printed in Europe, USA, Canada, Australia, Japan

Cover: Foto ©berggeist007 / pixelio.de

More available books at **www.hansebooks.com**

WITH ITS

MEDICO-LEGAL BEARING

—⚬—

A PRIZE ESSAY

Read before the Medical Society of the State of California,
April 17th, 1885

—⚬—

BY

WASHINGTON AYER, M.D.

Member of the "Harvard Club;" Ex-President of the Medical Society of the
State of California; Ex-President of the Medical Society of the City
and County of San Francisco; and Ex-President of the
San Francisco Medical Benevolent Society

—⚬—

SAN FRANCISCO:
W. S. Duncombe, Publisher
30 Post Street

DEDICATION.

TO THE HON. D. J. MURPHY,

JUDGE OF THE SUPERIOR COURT,

CRIMINAL DEPARTMENT OF SAN FRANCISCO,

THE FOLLOWING PAGES

ARE RESPECTFULLY DEDICATED

BY THE AUTHOR.

PREFACE.

In preparing this "essay" I have endeavored to show, by introducing various subjects connected with physics, that all mental alienations are dependent upon changes and developments of the brain, and not upon the mind itself as an essence emanating from the crucible of Omnipotent Will, to be measured and weighed as the weird offspring of unrelenting fate.

While not intending to criticise law as a science, the medico-legal bearing of the subject has led me to give such interpretation to special law, and criticise its application to specific cases, as would enable me in the most forcible manner to present reasons for the labor and burden of proof against the theory of mania transitoria I have undertaken.

Feeling that public sentiment has too often and too long been outraged by the admission of this theory into the pleadings for the benefit and protection of those who have most wantonly committed homicides, I have devoted my time to the investigation of the subject, with no motive beyond a wish to serve the interests of the public by endeavoring to make the murderer responsible for his crime, and to show that if legal penalties are not inflicted upon the guilty, all virtue and morality will soon be trodden under foot, and communism or anarchy will put at defiance the majesty of law.

I have cited several noted cases that have come into our courts, for the purpose of showing the importance of instituting a thorough investigation of this subject: and have referred to many facts connected with the changes that take place in the brain to prove that the mind is not of itself diseased, and cannot possibly become diseased; for the human mind is the individualized offspring of the Creative Mind, which gave birth to worlds and brought order out of chaos, and must forever remain unchanged. Whatever is subject to disease is also subject to decay and death; but the mind is a part of the Immortal Essence, which survives decay and will never die.

TRANSITORY MANIA

MEDICO-LEGAL BEARING.

Mr. President, and Fellows of the Medical Society of the State of California:

"The labor of life is a constant struggle between the acts of conscious volition and the automatic impulses of the emotional regions of our being," and as we are borne along upon the hurrying breath of time, how much we see to do, and yet how little we accomplish! While we pursue the study of the brain and search for the seat of mental activity, the shadows of ignorance and doubt give way to the fulfillment of prophetic knowledge, whose stream leads on to the "ocean of truth" and "ineffable light." We live in an age of investigation, an age in which *declaration*, unsupported by demonstrative knowledge, is of no value, and it is the right of every man to differ from his neighbor.

And now, in a spirit which cherishes love for justice, reverence for law, and sympathy for the unfortunate, I invite your attention to the subject of Mania Transitoria, with its medico-legal bearing.

Fully conscious of the many difficulties surrounding the intelligent discussion of this question, I shall not confine myself strictly to a synthetic line of argument, but shall call attention to such subjects as occur to me to be of importance to the profession in this connection; such as expert testimony, questions of law and rulings of courts upon questions of *irresistible impulse, emotional insanity*, and moral obligations to society, as the subject is of special legal importance, and involves many of the perplexities of jury trials. I shall also present such reflections as naturally arise as the outgrowth of my argument, without claiming for them original scientific conclusions, hoping to awaken some new thoughts in the mind of the careful student upon this subject.

With the results of the investigations of Brown-Sequard to guide us in our inquiries, as well as many other authors who have written upon the subject of nervous and mental diseases, and the researches of Luys upon the "Functions of the Brain," aided as they have been by experiments in vivisections, there can be no doubt of the localization of nervous disease and mentality within the cerebral hemispheres and medulla oblongata ; and the correctness of the theory of the proliferation of brain cells, and that they are the seat of mental activity and repositories of memory and knowledge, can hardly be doubted. With this view of the subject, and our knowledge of the functions of the brain in their control of all physical functions, we are led to conclude that the advantage one has over another in scholarly acquirements lies in his greater *receptivity* as the result of achievements in scholastic learning.

It is frequently asked how it occurs that one in advanced years remembers the things of childhood with more freshness than the things of yesterday. This can be explained upon the theory of the proliferation of brain cells as the seat of mental activity. To make the explanation simple, we will assume that the cells are never separated, but remain intact, and form a pyramid or cone, the base representing youth, and the apex, age. Now, as light falls upon the eye to produce an image upon the retina, or sound comes to the ear, the optic and auditory nerves vibrate and put these spiral columns in motion, and the greatest displacement will be at the base, the largest axis of spherical action, and the point where the first impressions of lisping childhood are made; each lobe and convolution of the brain, with its millions of cells, being assigned for special memories; and where the vibration is greatest the memory will be aroused first and strongest, while the effect upon the apex will be scarcely disturbed and soonest obliterated. Hence, the memory of childhood impressions are soonest brought before us in age.

As we are dealing with physical functions which control and enable us to classify mental activities, the discussion of mental disturbances justifies the introduction of this theory of *thought impressions* upon proliferated brain-cells as a basis of further investigation, and a means of obtaining experimental knowledge of the psychic laws which make man a responsible being.

Until within the past few years, brain-cells were considered only a "shapeless mass of protoplasm;" but now they are known to be delicately constructed, and possibly contain the realistic germs of life-forces which survive physical decay.

In this connection, as a matter of speculation, we may be justified in saying that within these microscopic cells, *thought impressions* are made, and there remain till interrupted by disease and decay, as impressions of objects are made upon the glass or other polished surfaces within the camera, to be again transferred ; or as sound is collected to be reproduced by the phonograph at the will of the operator; and it seems to me possible that men may yet learn that the elastic atmosphere is forever vibrating with vocal strains as an evidence of the realisms awaiting the intellectual enjoyments of etherealized matter, called mind. ·

In the reverberation of the clouds, in the tumult of the air, in the commotion of waves, in the summer calm, and wherever we turn, we find the expression of a living thought, begetting inspiration and urging man to investigate the causes of the wonders he beholds, and appropriate his knowledge to advance his happiness and comfort, and no obstacle seems too great to be overcome.

In the study of the mind and the effects of physical disturbances upon mental activity, we seem treading in the path of the "unknowable," and grow bewildered as we cautiously attempt to lift the veil that conceals all, to gain a glimpse of primeval causes. Though we cannot paint the dying refrain of summer upon the canvas with the woods and fields, resonant with song and redolent with the perfume of flowers, it will continue to live in the unknown recess of mind, to be reproduced again and again, and add to the pleasures of the future through the memory of the past. But there yet remains much we can learn in the vast fields of discovery, and can accomplish and know, what is now unknown and full of seeming mystery, awaiting the voice of science to be declared to man while he struggles amid the unmeasured forces of life in search of new truths, amazed in the contemplation of the wonderful works of a Creative Intelligence.

But we cannot think and speak much in advance of the times in which we live without some accusation of heresy being hurled at those who dare oppose the teachings of the present age, whether such teachings be in the line of legitimate medicine or opposed to popular theology, or the ethics and precepts of law ; yet an honest effort cannot be without its ultimate merited reward.

The mind grows weary and unsteady in its volitional manifest ations in abeyance to the laws of physical inertia, as a general proposition—exceptions to be noted—and the conscious volition then fails to direct the eliminating forces of intellectuality in the pathway of mental activity and reason, being governed by the same laws which apply to the disordered functions of physical life, dependent upon lesions of the brain and spinal cord, or medulla oblongata, as witnessed in myelitis and the various forms of paralysis ; and whenever the *pons varolli*, crus cerebri, corpora striata, or any of the nerve centres are over-stimulated or pressed upon, we shall invariably witness either a loss of motion and sensation, and loss of will power upon the parts affected at the same time, or an error of judgment ; and no one can reasonably doubt the intimate and inseparable relations of will and motion as a physiological proposition in all healthy bodies, except such motions as are or may be governed by their own specific ganglia, and those which are from their nature involuntary, as pulsation of the heart, nictation, respiration, paristole, etc.

In all the civil and social relations of life, where there is witnessed refined tastes and polished manners, we must acknowledge the importance of a clear mental capacity to appreciate in the selection of companionship. No question connected with medical science more deeply concerns the best interests of society and social life than that which is known as "*mental disease;*" yet how little is known by a majority of medical gentlemen upon so important a subject.

The literature upon this special department of scientific knowledge seems to be ample as a basis of investigation, to enable the analytical mind to make such delicate and clear distinction between *mens sana, aut non,* that, when we are called into criminal courts as experts, the learned members of the bar and the judges upon the bench may be answered intelligently and with credit to the profession we represent.

Works upon medical jurisprudence only partially treat upon this subject, presenting it in its medico-legal bearing, and leave the important functions of the brain and effects of molecular perturbation on mental forces to be investigated in another department of learning; and intelligent juries now look to the physician to aid them in determining the difference between an involuntary and unconscious action and a voluntary act. As a simple illustration, *winking* may be an involuntary action, while closing the eyes to slumber is a voluntary act. Walking, talking, and singing while asleep are *unconscious* voluntary acts, while the same exercises when awake constitute *conscious* voluntary acts; for in a healthy body the suspension of volition would be followed by a suspension of motion, as witnessed in paralysis. This is axiomatic.

J. H. Belfour Browne is generally considered authority upon forensic medicine, but a careful analysis of some of his statements will show he is not always correct. He says: "Dementia is an exaggerated enfeeblement of age, a more ruinous dotage. It is dependent upon exhaustion and torpor of the mind, that the mental house is in ruins, etc."—Page 276. This, however, is not correct, for dementia is the enfeeblement of the forces of mental activity, dependent upon cerebral disturbances of various kinds. Again he says: "Dementia is the inertia of rest—mania the inertia of motion."—Page 279. Now there is nothing in medicine nor in physics to support these statements, and give them any importance in a court of justice.

As intelligent testimony is the motor power applied to the machinery of law, which weaves facts into every conceivable fabric, and coloring to be presented for judicial inspection, and to juries to enable them to return just verdicts, if the physician would become a valuable witness, and assist criminal courts in the administration of wholesome laws, he should make himself familiar with the medico-legal bearing of mental disease upon criminal acts, before going upon the witness stand to give his testimony.

For the benefit of the younger members of the profession, I would say, we must always be on the *qui vive* while in court, as hypothetical questions are often asked, interwoven with paradoxical theories to test the competency of a witness, rather than to elicit any material facts connected with the subject of insanity.

To obviate the embarrassment that naturally follows upon an evasive or confused answer, a common estimate of the value of symptoms should be *formulated* and agreed upon, as the result of experience, observation and careful investigation.

At the present day a large number of the intelligent people of our country are inclined to look with disfavor upon all medical expert testimony before juries, and a majority seem to regard the proceedings of our criminal courts, in many cases, as farcical dramas brought upon the judicial stage, before the bar of justice in opposition to the spirit of the higher codes of law, which inflicts punishment upon the guilty; and justice stands behind the proscenium taking no part in the proceedings, and leaving the criminal with his weakness and errors to the sympathy of juries, often composed of men incompetent to tell the difference between a hypothetical question and an axiom of law. This is especially the case when the criminal is being tried for homicide, and the defense puts in the plea of irresponsibility, on the ground of transitory or emotional insanity. Who is responsible for the failure to convict in such cases but the medical expert, and those who mis-apply and mis-interpret facts? There are but few eminent writers upon this subject who give unqualified endorsement to the theory of transitory mania, yet the sophistry of learned attorneys and the testimony of pliant medical witnesses, carry conviction to the minds of jurors that the murderer is not responsible under such alleged conditions for his act.

This was illustrated in the suit of the People *vs.* Laura D. Fair, one of the most noted cases that appear upon the records of our criminal courts of the State of California, where the plea for the defendant was transitory or emotional insanity, "and the act was non-volitional." In this case the effect upon the jury of the testimony of the physicians who were called as experts on the first trial, was not such as to receive any favorable comment, as it did not appear to be wholly in accordance with the facts sought to be presented; nor was it free from the appearance of *bias*, a condition of mind most unfortunate for a witness in a court of justice. On the second trial the effect of expert testimony on the part of the physicians was such that it could not be overcome by the eloquence of polished rhetoric, finished oratory, and scholarly argument, and the prisoner was acquitted.

MURDER OF CHAS. DE YOUNG.

Nothing could be more injurious to church ethics, or more insulting to Christian precepts, and more demoralizing to the youth of a populous city, than the example presented by the killing of Mr. Chas. De Young, by Kalloch, a minister of the gospel.

In this case we find the law was made a creature of sympathy by a maudlin sentiment that echoed from the bells upon the church towers along the corridors and through the aisles of the temples of worship, until it reached the ears of the jury and deprived justice of its executive authority; and a greater outrage upon offended law can scarcely be conceived than was witnessed in the result of this trial. Here the expert testimony was much more declamatory than logical, and was in no wise entitled to the position it occupied in the category of evidence for the defendant.

The great effort of prominent members of the bar appears to be to protect the criminal against the law, rather than secure the enforcement of its penalties, and this is the *hoc opus* which tends to make crime rampant and subvert good government. Experience and daily observation teach us that it needs no forensic oratory, no long recitations from Shakespear, no quotations from Milton's Paradise 'Lost, no quotations from Young's Night Thoughts, whose windows of light were forever darkened, or from Cowper, to inform a jury that

"Unnumbered throngs on every side are seen,
Of bodies changed by various forms of spleen,"

and that the prisoner charged with homicide is not guilty, provided *handy* witnesses are secured, and the *right* kind of men are placed in the jury-box.

You may ask, "What have medical gentlemen to do with this?" Let me assure you we have much to do with the conviction of the guilty, by giving testimony, supported by the facts presented by the laws of physiology and psychosis, which dignify and control human action.

The study of nervous diseases and mentality, and inquiries into the forces governing the functions of the brain, by those who have had large opportunities for experimenting upon the lower animals in vivisections, have called in exercise new reflections, and given new encouragement to the investigation of mental diseases. And while mind cannot be examined, only by logical deductions, much that controls its functions can yet be learned, as we learn the nature of imponderable agents, by their effect upon animate and inanimate objects, as reason lifts the veil beneath which volition dwells, and discloses the working of the forces that direct and control the acts of men, and define his responsibility in every individual act.

It is a maxim in common law that "all persons are presumed to be innocent until proven guilty." Judge S. H. Dwinell, in his charge to the jury in the case of the People *vs.* Laura D. Fair, says: "All the presumptions of law, independent of evidence,

are in favor of innocence, and every person is presumed to be innocent of crime until he is proven to be guilty." But the law really does no such thing in its application; if it did, it would not arm its officers with authority to arrest and hold persons *in duress* until such time as they may be brought into court to prove their innocence. Otherwise the rights of individuals would be taken away, without any possible redress at law for inconvenience and hardships endured; for no person can justly be restrained of his liberty unless he is presumed to be guilty or insane—except he be restrained as a witness in the interest of justice—and officers making arrests would be liable to be charged with being governed by an improper influence—*malo animo*—in the discharge of their duties, and this would be a conspicuous error.

In dealing with crime, "Justice *a priori* ascribes responsibility to all who commit it," and consequently must look with suspicion upon all placed under arrest, whether sane or insane, and the condition of the mind in its capability to determine right from wrong must alone decide the responsibility in any individual case, for ignorance of law is no excuse for wrong doing. *Ignorantia legis neminem excusat.*

When the maxim referred to is the ruling of the court, for the purpose of arresting a popular verdict in any case before it has been tried, it is well enough as tending to prevent meddlesome interference with the proceedings of the trial. But we have no dealings with maxims in the abstract which apply to the guilt or innocence of any party; and all the court and jury want us to do is to make a clear statement of facts relating to mental conditions, and nothing more in cases of alleged insanity.

We must not only have ideas, but must have positive facts upon which to base our testimony.

AN IDEA.—An idea is an element or condition of mind as it relates to sanity or insanity, and develops rapidly into various expressions of language and acts, which may be rational or not, according to the varying circumstances upon which it is dependent.

IMPULSE.—Impulse is a sudden feeling different from that which is controlling the action of the individual at the time it occurs; and while it is not voluntary, and cannot be brought into play by any force of the mind, it is subjective to the will when manifested; and this is in accordance with the experiences of daily life. It differs from an impression in this: An impression is the consciousness of the existence of something, while an impulse relates to an action or desire to act.

Impulse is not the product of thought which springs from or enters into the domain of mind, but is developed by some extraneous or concentric action of the *vital aura*, or surrounding erethism of the body; while many of the moral feelings are the offsprings of thought, such as sorrow or joy, and sometimes *thought*—very emphatic—is the offspring of pain, as experienced in gout or neuralgia, and this paradoxical condition is dependent upon the flex and reflex nervous action, stimulating certain cerebral nerve centers. But "impulse" is exhausted the moment it is felt, and gives birth to reflection, which urges one on to *uncontrolled* action; but never irresistible in its nature, as reflection is always rational in character, and must be subjective to the will.

Upon this is based all of the theory of transitory mania, which is only a convenient myth—a huge joke on ethics and criminal law—merely a word-structure of defense without a possible reality; and the learned members of the bar do not believe in its existence.

A more ridiculous burlesque upon justice cannot well disgrace the procedure of criminal courts than is witnessed in a trial when the argument for the defense is based alone upon this theory; for it is not within the reach of psychic reasoning, or the logic of presumption, and is little more than nonsense, imbeded in metaphysics, without a ray of intelligent possibility to dignify an argument in its favor, and is far more quixotic than the Rosicrucian philosophy, or the animisms of Stahl.

This theory not only aims to encourage crime, and relieve the criminal from responsibility, but lessens public respect for the Temple of Justice.

And in this we see justice struggling—

> "See physic beg the Stagyrites' defense—
> See metaphysics call for aid on sense."

With this view of the subject, I am conscious of invading the domain of cherished opinions of some of the eminent writers and opposing their theories, yet I do so not in a spirit of hostility, but with a desire to advance justice and the interests of the profession.

> Oh, common sense, divinest child of earth !
> May reason's choral voice thy praise prolong,
> .Till nature, wearied, sinks beneath the song.

Judge Hoffman's opinion of emotional sanity and transitory mania.—In the case of Mary Jane Sweeney, the defense begged leniency in consideration of Sweeney's emotional insanity. The Judge replied : " This plea of emotional insanity, or transitory mania, or whatever name the excuse may be given, has become almost ridiculous." " Our experience in California in respect to this subject has led us to regard the present aspect which the insanity plea has assumed, as repulsive to justice and fatal to society."

In further support of the statement that the learned members of the bar have no confidence in the correctness of this theory, I will call your attention to the following remarks of Prosecuting

Attorney J. N. E. Wilson, in a case recently tried in one of our Superior Courts, of the People *v.* Kennedy. He says : " Rumor has it that defendant's counsel possesses one of the finest medical libraries on the coast. Rumor also says that the honorable gentleman boxed it up and stowed it away when he commenced this case, because he was fully aware that no such thing as 'emotional insanity' could be found in it. But he goes on with the case, and instead of calling it 'emotional insanity,' gives it the title of 'disordered mental design.' " This satire would probably have been enjoyed by the learned attorney had his position been reversed in this case, as on a former occasion he had evoked an adverse decision from the Supreme Court upon a question involving a similar theory.

In the 62, California Reports, page 123, in the case of the People *vs.* T. J. Hein, Judge McKinstry, in giving his decision, says : "It will be seen that the English courts have refused to recognize the co-existence of an impulse *absolutely* irresistible, with capacity to distinguish between right and wrong with reference to the act. It cannot be said to be irresistible because not resisted. Whatever may be the abstract truth, the law has never recognized an impulse as uncontrollable, which yet leaves the reasoning powers, including the capacity to appreciate the nature and quality of the particular act, unaffected by mental disease. No different rule has been adopted by American courts." This was concurred in by Morrison, C.-J., Myrick, Sharpstein, Ross, and McKee. J.J., and was a complete answer to a question asked by Judge Darwin pending the trial.

The New York courts hold to the same doctrine, as appears in the New York Reports in the Court of Appeals, page 469, in the case of Mark Flannagan, Plaintiff in Error *v.* the People: " In this case all the judges except one concurred in the opinion of Judge Tindall, C. J., and the case is of the highest authority ; and the rule declared in it has been adhered to by the English courts."

IRRESISTIBLE IMPULSE ILLUSTRATED.

When the staging gives way, the irresistible force of gravitation causes the laborer to fall to the ground ; and when this force is overcome by inflating a balloon with a much lighter air than that which surrounds the earth, those who leave in the basket prepared for an aerial flight, are *irresistibly* borne far away among the clouds. In one case the irresistible force is a fixed law of nature ; the other simply a relative force, subjective to chemistry and mechanics, and is a part of experimental life. But by taking the necessary precaution, the voluntary danger and the involuntary act of falling may both be averted. This simile will apply to the so-called *irresistible* impulsive acts of men, which are no more than *unresisted* acts, which may be avoided. In one case the act is voluntary, in the other involuntary ; but if the party falling is injured, he is responsible in the one case by his folly, in the other by his carelessness, and cannot escape responsibility in either ; for vigilance and prudence would have enabled him to remain unharmed, since he could not be raised into the basket nor lifted upon the staging by any irresistible force, while knowing the act he was performing (idiots always excepted).. If a man is seized with a spasm of anger and kills his brother, he must be held responsible, and should be sent to one of the lunatic asylums for life, for he voluntarily steps into the basket to be borne away by the balloon, inflated with rage ; or, his voluntary *involition* causes him to fall.

It would be a dangerous doctrine to establish, to say a person was competent to determine right from wrong in any particular act, and at the same time was impelled by an irresistible foe, unseen and unappreciable, which was urging him on, *vis a tergo*, to commit a crime which he was capable of appreciating, and realizing the consequence of its commission, and yet was not responsible for the act.

No man arms himself with a pistol or knife and visits the house of another party whom he suddenly kills, without knowing what he is doing, and what he intended to do before leaving on his mission of crime, unless he was insane before and remains insane after the act.

Alfred Swain Taylor, in his work on Medical Jurisprudence, says: "The main character of insanity in a legal view, is said to be the existence of a delusion, *i. e.*, that a person should believe something to exist which does not exist, and that he should act upon this belief."

But if the theory of mania transitoria be correct, the party so attacked does not have time to indulge even in a belief of a delusion, but steps out of himself for a moment to give the body an opportunity to act and kill some one, and then steps back again just in time to take the body away in a perfectly healthy condition, uninjured by the sudden transition, from a subjective agent to an independent actor and *vice versa*. Now can anything be more absurd? Yet many intelligent medical gentlemen have been upon the witness stand as experts, and testified before courts and juries to the existence of this mental condition— this *psychic ledgerdemain* which is beyond the reach of mental philosophy to explain.

Such testimony is the *opprobrium medicorum* of the profession, while the atmosphere is redolent with inquisitive objections to such assumptions of learning. The highest medical authority has never undertaken to prove by any known physiological or psychic laws, how such a mental condition can possibly have an existence as transitory mania.

But in a strictly medical view, insanity does not exist in the mind *per se*, for that is not *ipse facto* diseased, but is dependent for its aberrations upon some abnormal condition of the physical man, which may be either organic or functional, and usually found in a change in the structure of the brain and the surrounding tissues. By reflex action the erethism of remote organs may be conveyed to the brain, over-stimulating the whole mass, or certain ganglia corresponding to the seat of nervous activity, which supplies the organs effected with their sentient and motive sensibility, or automatic consciousness. And often the dyscrasia of the party, without any localized morbid change, effects the *vital aura* and periphery of the entire nervous system, producing hyperæmia of the brain, followed by *disordered mental manifestations.*

As before stated, the mind is not, *ipse facto*, diseased. Here, I apprehend, lies much of the error and perplexity experienced by witnesses who attempt to define a disease of something that exists but has no length or breadth, the same as they would define the disease of the atmosphere, which is rendered *toxic* by the introduction of noxious gases during respiration of animals or vegetation, or from the exhalations of forests, or other causes, by which the oxygen is displaced and a new compound formed, that enters the lungs; or by heat, which lessens the volume of oxygen without changing its ratio as the air becomes rarified and expanded. Such a course of logic cannot be maintained, for the elements of mind are not tangible, like the odor of flowers. The difficulty of *understanding* this question and not being *understood*, is in consequence of not recognizing the truth of the adage, *mens sana in corpore sano.*

Cases of insanity following the use of alcohol or opium are always preceded by *delirium*, showing that changes do take place in the brain from over-excitement, until a permanent lesion is formed and becomes localized. The same may be said of insanity arising from cerebro-typhoid fever, supervening upon disappointment and business losses, all of which are the psychic manifestations of physical disease.

After carefully studying the functions of the mind, which render it voluntary or involuntary, I fail to find any evidence to prove the existence of *transitory mania*, beyond the declarations of the criminals themselves.

Such opinions, which have become quite too popular for the public good, and upon which often rest verdicts of juries, appear to have been formed and agreed upon to meet certain cases after deeds of violence had been committed, and are purely *ex post facto*. When there is a temporary suspension of the conscious action of the will, affecting any of the moral sentiments, no harm can possibly be done; for the consciousness of an idea to act is lost. And when the suspension of the will-power, from whatever cause, affects the physical forces, motion is irregular, and seldom in a direct line of action; the motor nerves always manifesting a volitional disturbance, as witnessed in *paralysis agitans*, and it is not possible for the will to be suspended in the middle of an *arc*, being described in the act of a blow and before its completion, and again suddenly regain its full force the moment the act is performed. Such doctrine would be dangerous, and, if maintained, would render our criminal courts powerless to administer justice, and the physician should be held largely responsible for encouraging it.

In this theory we find all that supports the argument in favor of *transitory* mania, which in a spirit of great generosity should be considered only as *transitory cussedness*, in distinction to other vicious traits, and differing from moral depravity in many of the essentials by which man is recognized as being endowed with a moral and physical nature.

We are told that " charity is kind and of long suffering," and that it is better that "ten guilty persons should go free than that one innocent party should suffer." This, as an expression of a Christian mind, and in accordance with church ethics, is well enough; but when we devise some theory by which the guilty may go unpunished, we encourage the infraction of law and degrade public morals, which is not in accordance with the teachings of the gospel of Christ, or any sentiment of morality.

Whatever theories are advanced in regard to the insane, there must survive a consciousness of some sentiment in the mind of every intelligent person, and a feeling of regret for the wrongs of the aggressor, which leads one to desire to remedy all within human power that is wrong ; and to be just, we must remember the offended have rights that must be respected, and we must be prepared to prove the existence of some conscious reality behind every act, as well as that which follows a crime, and *vice versa.*

As we become more familiar with the social influences of life, which shape the course of individual action, the more we are impressed that life itself is a mimicry as related to the daily affairs in which we are most interested; and it is of the most profound importance that we should make ourselves familiar with all the possible details of pathology, that we may become competent judges of the functions of health and disease, and be able to determine their influence over the moral and mental forces which determine the nature of human action. Here is where we can maintain an honorable distinction above the average witness, and render valuable service to an honorable and learned profession in the halls of justice, in the interest of public morals and human safety.

What is involuntary action, and what emotional? Involuntary action is witnessed in respiration, nictation, pulsations of the heart and spasms of tetanus. Emotional action is seen in weeping, sighing, and laughing. The difference is found in the fact that the former acts are without any conscious effort of the will :

while the latter acts are dependent upon impressions made upon the mind by surrounding influences, as reading, seeing, or hearing some sad or mirthful story.   And while these manifestations cannot be wholly controlled by the will at the moment of occurrence, they are not involuntary, since the volition is active and conveys impressions to the seat of the mind, which arouse reflections of pleasures or pain.

If the prism through which the rays of light pass is defective, there will not be a homogeneity of colors produced, and this condition will confuse and antagonize the science of dioptrics, and of the solar spectrum in the simple multiple rays or factors, which unify the rays of light around us, or separate them into primitive colors; and this theory holds good in the various cognate branches of science, and in mental pathology.

We do not see light, but realize its presence, and recognize distances by the eye; nor do we see mind, but recognize its force by the effect it has upon others in the silent language of invention, and social and political leadership, and realize its changes by the tone of the voice, the laugh, the smile, the tears or sighs; and can read the unwritten law, governing mind by voluntary action, as we judge of heat and magnetism by their effect upon vegetable and animal life.   All disorders of the mind are so many evidences of molecular perturbations of physical forces, dependent upon some positive lesions of the anatomical structure of the brain, or of those organs and functions which exercise a strong controlling influence over the same.   Hence we may reasonably infer there can be no insanity arising from the mind *per se*, beyond its influence upon some of the physical functions; and all the manifestations of anger, hate, frenzy and impulse, or emotion of whatever nature, are volitional, and subjective to a proper exercise and control of the will, and man is responsible for their consequences.

DUTY OF PHYSICIANS.

It is as much the province of the physician to endeavor to have the criminally insane taken care of, as to advise measures for the care of those who are incapacitated from any of the casualities of life, and I would recommend that the committee of this society on legislation be especially instructed to endeavor to secure the passage of an act, to provide for the care of the insane criminal in a separate building from the morally insane. I would also recommend that the law of Massachusetts upon this subject shall form the basis of such legislation.

The law reads.as follows, viz:

SEC. 20.—"When a person indicted for murder or manslaughter is acquitted by the jury by reason of insanity, the court shall cause such person to be committed to one of the State lunatic hospitals during his natural life."

SEC. 21.—"Any person committed to a State lunatic hospital under the foregoing section may be discharged therefrom by the Governor, by and with the consent of the Council, when he is satisfied that such person may be discharged without danger to others."—*General Statutes, chapter 214.*

In kleptomania there can be no real motive to wrong another or acquire the stolen property, for motive must be prompted by an idea of revenge, or some possible advantage, and things are often stolen which can be of no possible benefit to the parties who steal. In pyromania the party applies the torch to the building without a motive to harm another. In the former case the stolen goods are concealed, and in the latter the property is destroyed, and no benefit is to be gained by the commission of crime to either party, nor is there any fixed motive for the crime.

But the absence of a motive to be benefited or to injure another does not make the offense less punishable under common law by placing the parties under restraint.

Then, admitting the theory of transitory mania to be correct, and that homicides are committed without a motive of gain or injury to another, and the absurdity of the law that punishes the former and acquits the latter is readily apparent, and the theory of transitory mania is made to appear ridiculous.

Taylor (page 674) remarks: "It cannot be denied that the doctrine of 'irresistible impulse' has been strained in recent times to such a degree as to create a justifiable distrust of medical evidence on these occasions."

"It is obviously easy to convert this into a plea for the extenuation of all kinds of crime for which motives are not apparent, and thus medical witnesses often expose themselves to severe rebuke. They are certainly not justified in setting up such a defence unless they are prepared to draw a clear and common-sense distinction between impulses which are *unresisted* and those which are *irresistible*."

This *irresistible* theory would deprive a man of his individuality, and make him a frail instrument in the hands of an unknown foe, and subvert public morals, pervert the purest principles of law, and endanger human life.

Maudsley says: "Many cases of the so-called *transitory mania* are really cases of mental epilepsy," and cites a case of a patient in the French Asylum at Avignon. A similar case is related by Esquirol of a Swabian peasant who killed his mother (page 234).

Falret describes epileptic vertigo as a sort of *petit mal*, or transitory disease, with no pathology : "These peculiar states of epileptic consciousness are not only of great psychological interest, but also of practical consequence in relation to the question of responsibility, for it is obvious that deeds might be done by an individual when in an anomalous state of consciousness, of which he might have no remembrance when in his really normal state, and for which he could not justly be held responsible."— *Maudsley*, page 238.

It seems from the foregoing that the author considers *transitory mania* to be of an epileptic nature entirely. From this view of the question I do not dissent, it being one of the paroxysmal manifestations of a disease known to exist, as delirium is the result of fever and hyper-stimulation of the brain in intoxication, and nothing more.

In all such cases there are prodromata, or physical conditions existing, which affect the normal status of the will, and which antedate acts of violence; and those thus affected do not find their remedy and cure in the act itself, as has been often alleged by those who desire to acquit the homicide upon the plea of transitory mania, and who declare the actor is instantly restored to consciousness after a deed of violence is committed.

J. H. Balfour Browne, page 170, says: "Again, such a disease as transitory mania, mania which suddenly appears and suddenly disappears, is, to our thinking, an impossibility."

Dr. Hammond remarks : "The doctrine that an individual can be entirely sane immediately before and after any particular act, and yet insane at the instant the act was committed, is contrary to every principle of sound psychological science."

Even in the most striking instances of what is called transitory mania or morbid impulse, the evidence of pre-existent and subsequent disease of the brain will be found if looked for with diligence and intelligence.

Dr. Gray, a distinguished authori y, says : "I am not going to deny the existence of transitory paroxysms *in* insanity, either in epilepsy or in the frenzy of melancholia, or in ordinary cases of insanity where paroxysms suddenly arise and suddenly disappear; but until I have seen more than I have yet seen, and until I have read something more authentic than I have yet read, I must fail to see insanity in any case which arises when the premonitory symptoms of the disease run the rapid course of a few minutes, when the person commits a crime and then is well."

E. C. Spitzka, in his work on Insanity, published 1883, page 154, says: "Numerous instances are recorded where persons, previously of sound mental health, have suddenly broken out in a blind fury or *confused delirium*, which, passing away in a few minutes or hours, left the subject deprived of a clear, or any, recollection of the morbid period, and generally concluded with a deep sleep." These conditions are witnessed in cases of concussions of the brain, when a person remains unconscious for a longer or shorter period, and upon being restored to consciousness has no recollection of what has occurred, with the exception that no fury is manifested.

This author says "observers designate this condition as transitory mania." "Others term it transitory melancholia, and others class it among epileptic disorders." But to our thinking it should be classed with hysteria, dependent upon the inhibition of the nerve centres and reflex action.

The same author says, page 155: "But it would have to be considered a remarkable form of epilepsy in which there was but a single epileptic attack." "Transitory mania, or frenzy, is a comparatively rare affection, so rare that many asylum physicians have never seen a case of it; the writer has likewise never had that fortune." According to the same author, " Foville, in his 'Annales Medico-Psychologiques' of 1874, declares moral insanity and mania transitory false, absurd, ridiculous, and above all, unworthy of being received by the courts."

Cook, another author, claims that " transitory mania is a cerebral epilepsy."

Kinnon says: "You cannot prove the epilepsy; you can prove the mania, and it is transient;" "and is it not as easy to accept the theory of transitory mania as it is to go wandering after a far-fetched forced explanation?" In reply I would say, it is easier to accept this declaration than to undertake the labor

of proving or disproving its correctness; but it is not in accordance with the spirit of fair inquiry to admit anything that is not proven, for no axioms upon these questions are presented for our guidance that have not been tested in the crucible of science. A simple declaration of the existence of anything does not establish a fact, any more than the declaration of an action constitutes an act. We are not prompted by any spasm of curiosity or aggressiveness, but are investigating this subject in a spirit of great kindness and sympathy for the unfortunate, and that we may be better prepared to assist the courts in the just administration of the law when called upon, when the plea of the defence is transitory mania ; for now there seems to be a greater effort to shield the guilty than to protect the innocent.

From our knowledge of the functions of the brain and the media through which psychic and volitional forces pass, we have before us evidence to show that outward manifestations of violence are in consequence of a disturbance of the vaso-motor centres which lie along the floor of the fourth ventricle, and are imbedded·in the gray corticle of the spinal cord, and from an over excitation of all or any of the cerebral ganglia, and whatever disturbs the substance of the brain.

The daily panorama of the changing scenes of life is forever making impressions upon the mind for pleasure or pain, as the strong rays of the sun produce photophobia or give pleasure as one gazes upon the scenes of an outspreading landscape. Or the brain may become so acted upon that the mind will lose its power of discriminating judgment of colors, forms, and objects, through the defective medium of sight, occasioned by straining the eye, or in astigmatism, or double dyplopia, when errors of vision become provocative causes of mental alienation, as the optic lobes are falsely stimulated by the imperfect image upon the retina coming through the refractive media of an astigmatic eye.

This effect is not wholly confined to the optic lobes. Stimuli, if sufficiently strong, applied to the afferant nerves, will inhibit, *i. e.*, " will retard or even wholly prevent reflex action " (Pastor, page 419). These'facts may be applied to the reflex action of psychic forces which awaken another train of errors in judgment, and can only be overcome by a careful course of reasoning. But an error of judgment, however persistently followed, must not be received as an evidence of insanity.

DELUSION AND HALLUCINATION.

Delusion is a deception as regards the existence of truths. Hallucination is a deception as regards the existence of things. The former relates to abstract, the latter to concrete subjects. For instance, I am told of the existence of a great conflagration, and believe it, but afterwards I find it did not exist and that I was *deluded;* again, I think I see a conflagration and repeat my impressions of its magnitude, but afterwards find it did not occur, and then learn I was laboring under an hallucination.

All that appeals to reason and judgment through mental activity alone, that is not true, is a delusion—all that is presented to the mind through physical senses, that is not true, is hallucination.

Delusions arise frequently from physical causes, as one with an astigmatic eye, which is unassisted by proper lens, regards all round objects or circles as oblong ; also, where there is an imperfect formation of the *membrana tympani*, sound awakens an error of judgment as to its intensity and kind. When odoriferous particles fall upon the olfactory epithelium the sensation of smell is produced; but if this membrane be diseased, there will be an error conveyed to the sensorium, the same as occurs when there is a deformity of the retina in the objective sphere of vision.

These are some of the external and concentric causes of error of judgment, and serve to illustrate how disease or over-excitation of the cerebral nerve centres produce insanity or strong emotional feelings, according to the various media through which the vibratory excitement passes, but do not show where responsibility ceases.

Who cannot recall the memory of childhood with its scenes of pleasure or fear of parental discipline. At will, the whole panorama of life, with its daily etchings and embellishments, is brought before us to be with its original thought again compressed amid the subtle forces of an undefined existence; and we search nowhere but the brain for the forces which give character and direction to all the affairs of life, and hold all in reserve for the use of memory.

Physiologists distinguish two kinds of nervous action; one initiatory, the other inhibitory—the one originating, the other controlling. Now, just as the originating centres may be strengthened by indulgence, so may the inhibitory be made stronger by habit; hence, a man in ordinary health may be tempted by some false inducement to act, but he does not lose his power to resist the action. So a man may be tempted while suffering from disease, by some unreal object, some delusional belief, but it does not thereby follow that he is deprived of the volitional power of control over his acts, and that he is irresponsible for what he does on this account; for the originating thought is the force that controls all subsequent action.

" It is the feelings that reveal the genuine nature of an individual and the nature of his acts; it is from the depths of one's inner nature that the impulses of action come, while the intellect guides and controls; and accordingly in a perversion of the effective life is revealed a fundamental disorder, which will be exhibited in acts rather than in words."—*Reynolds*, page 592.

And here we find a ready solution to the problem of psychic forces, which prompt and lead on to *unresisted* criminal action, the will relaxing its hold upon fortitude and right; but not because the *unseen* is an *irresistible* foe that compels the unfortunate to the commission of crime, but by reason of the failure to exercise the moral faculties which *creative energy* has bestowed upon man.

Not an author which I have consulted has attempted to prove the pathology of transitory mania, or given any psychic or physiological reasons for their conclusions; the *ipse dixit* of the individual said to be so affected, is the only evidence given in support of the theory. And certainly this cannot be considered of any value in a court of justice; and it is surprising that any of the eminent writers upon this subject should attach so much importance to the declarations of parties judged to be insane but a moment before and at the time of a criminal act. I do not refer to the spasmodic homicidal impulse of known epileptics, or of the known insane, but have special reference to the *mushroom* development and decay of a *sui generis* type of insanity called transitory mania.

Wharton and Stille, page 710, say: "Mania transitoria is a sudden insane frenzy." 'As frenzy disconnected with physical suffering can have no possible pathology, *per se*, it cannot properly be held to be insanity, and should not be offered as an extenuating excuse for crime; for, being considered as a purely mental disturbance, it is only an increased state of *unresisted* passion.

Chitty, Forsyth, and J. T. N. Fontblanque make no mention of transitory mania.

Allan McLane Hamilton, physician to the Insane Asylum of New York, says (page 209): "When a crime is not to be accounted for, and completely inconsistent with the antecedents of one who is not known to be epileptic or insane, and when it is accomplished in a moment of fury, then we should examine whether these are aborted or nocturnal attacks of epilepsy."

"*Maniacal* rage of short duration is often epileptic in character, and its true character is often mistaken."

This author does not treat of transitory mania beyond this epileptic form of disease, and is wisely cautious about admitting its possibility, rendering his opinion of doubtful value upon any medico-legal question.

Ray, in his work entitled "Contributions to Medical Pathology," page 259, in the case of Bernard Congley, says : "It must have been a paroxysm of transitory mania, suddenly beginning and as suddenly ending, after the briefest possible duration. The cases of this kind of *mania* on record, though few, certainly are so well attested that we can scarcely deny the existence of the form of insanity which they illustrate. And it is a noticeable feature of most of them that the patient is bent on destroying life."   .

If bent upon committing murder, that fact implies the exercise of the will to accomplish some specific object; consequently his theory, if analyzed, would scarcely bear the crucial test, so as to be entitled to any more importance to the profession than the declaration of some less learned gentleman.

REPORT OF TRIAL OF SAMUEL M. ANDREWS, BEFORE THE
SUPREME COURT OF MASSACHUSETTS.

Dr. Edward Jarvis (page 173) testified : "Sudden manias vary ; sometimes they commence and terminate in a violent outbreak. They may come suddenly and cease as abruptly."

In the same case, page 187, George H. Choate testified : "I have had about 3,600 cases under my charge." "I have never known a case of insanity originating and terminating in a single act of violence. I don't believe such a case exists."

Page 188 : "There is a moment when insanity begins. There is a gradual increase of symptoms, and it does not reach unconsciousness without increasing symptoms."

Here we have directly opposite opinions from two eminent medical gentlemen, the one having the greater advantage over the other by his superior opportunities for observation, and he positively denies the *possible existence* of transitory mania. But, like all other authorities upon insanity, they make no attempt to prove, by any method of reasoning, why this condition may or cannot exist as a sequence of functional or organic disturbances. This is a part of the labor I have undertaken in preparing this paper.

In order to obtain an expression upon this subject from those best qualified to judge, I addressed a "circular letter" to the Superintendents of all the Insane Asylums of our country, and requested them to answer the following interrogatories, viz :

1. How long have you been connected with any institution for the treatment of the insane?
2. Have you ever seen a case of transitory mania that was not dependent upon some form of insanity, and that did not present itself as a manifestation of previously existing disease?
3. Do you consider it possible for transitory mania to occur as an idiopathic disease?
4. How many insane persons have you had under your care?
Remarks.

To these questions I have received the following replies. To avoid repetition I will place the answers in the order of 1, 2, 3 and 4, as the questions were given :

1—Seven years. 2—I do not recollect such a case. 3—Possible ; not probable. 4—650. Remarks : I have been of the opinion that transitory mania may occur, but in very exceptional cases.—*H. Wardner, M.D., Supt. Hospital for the Insane, Anna, Ill.*

1—Eleven years. 2—Not one. 3—Think not. 4—4,946. Remarks : I have also seen 25,000 patients in other asylums, but no case of transitory mania.—*E. T. Wilkins, M.D., Resident Physician of Napa (California) State Asylum for the Insane.*

1—Two years. 2—No. 3—No. 4—3,787.—*W. B. Fletcher, M.D.,
Insane Asylum, Indianapolis, Ind.*

1—Fifteen years. 2—No. 3—No. 4—2,000.—*Chas. P. MacDonald,
Supt. State Asylum for Insane Criminals, Auburn, N. Y.*

1—Seven years. 2—Never. 4—1,500. Remarks: Transitory mania
I think is of rare occurrence ; that it does sometimes occur I think there
is no doubt.—*C. W. King, M.D., Dayton, Ohio.*

1—Nine years. 2—Have not. 3—Do not. 4—2,000. Remarks : I do
not believe mania transitoria exists, *per se.*—*Randolph Parksdale, M.D.,
Supt. Insane Asylum, Petersburgh, Va.*

1—Twelve years. 2—No. 4—Several thousand.—*D. M. Wise, Supt
Willard Asylum for the Insane, N. Y.*

1—Sixteen years. 2—Most assuredly not. 3—No. 4—3,000. Remarks : Such a proposition as transitory mania is irrational, absurd, and
opposed to every theory of advanced psychistry.—*E. A. Kilburn, M.D.,
Med. Supt. Insane Hospital, Elgin, Ill.*

1—Twenty years. 2—Never. 4—7,000. Remarks : To be insane
there must be actual disease of the brain, which is not transitory.—*H. A
Gilman, M.D., Supt. Insane Asylum, Mt. Pleasant, Iowa.*

1—Ten years. 2—No. 3—No. 4—200. Remarks : I fully concur
with your views on the subject.—*R. C. Chenault, M.D., Med. Supt. E.
V. L. Asylum, Lexington, Ky.*

1—Sixteen years. 2—Never. 3—No. 4—3,602.—*John W. Ward,
M.D., Trenton, N. J.*

1—Eleven years. 2—No. 3—No. 4—1,297.—*J. W. Jones, Supt. Insane Asylum of Louisiana.*

1—Twenty-five years. 2—No. 4—2,500. Remarks : While my experience, as shown above, is against the existence of what you call mania
transitoria, my views of the nature of mind make such a condition scien-
tifically possible.—*P. Bryce, M.D., Tuscalusa, Ala.*

1—Eight years. 2—No. 4—5,000. Remarks : I have no reason to
doubt the authenticity of some reported cases of transitory mania. Mania transitoria differs somewhat as treated by different authors, and the

name does not sufficiently explain what is meant.—*W. B. Goldsmith, M.D., Danvers, Mass., State Asylum.*

1—Thirty years. 4—Several thousand. Remarks : Transitory mania as an idiopathic disease is not probable, judging from experience. I have never seen a case of so-called transitory mania.—*John B. Chapin, M.D., Supt. Hospital for Insane, Phil.*

1—Seven years. 2—No. 3—No. 4—699. Remarks : Daily average never less than 660. Your position is correct.—*C. A. Miller, M.D., Supt. Insane Asylum, Carthage, Ohio.*

1—Sixteen years. 2—No. 4—7,000.—*R. M. Wigginton, M.D., Supt. State Asylum, Winnebago, Wis.*

1—Eleven years. 2—No. 3—No. 4—2,500. Remarks : I do not believe there is such a condition as transitory mania, and think the use of the term should be abandoned.—*G. H. Hill, M.D., Supt. of Insane Asylum, Independence, Iowa.*

1—Thirteen years. 2—No. 4—2,700. Remarks : My judgment in suspense ; no such cases come to asylums.—*Richard Dewey, Kankakee, Ill.*

1—Twenty-seven years. 2—No. 3—No. 4—About 6,000. Remarks : I think it possible that there might be temporary mental aberration ; should not call it disease, but functional disturbance.—*C. K. Bartlett, M.D., Supt. Minnesota Hospital for Insane, St. Peters, Minn.*

1—Two years. 2—No. 3—I do not. 4—Some thousands. Remarks : When any well-marked neurosis, particularly epilepsy, can be shown, it would be difficult to disprove transitory mania.—*W. H. Mays, M.D., Assistant Physician State Asylum, Stockton, Cal.*

Remarks : Have never seen a case, and I do not believe in the theory of transitory mania.—*S. H. Talcott, M.D., Middleton, N. Y.*

1—Fifteen years. 2—No. 3—No. 4—Some thousands.—*James D. Moncure, M.D., Supt. Insane Asylum, Williamsburgh, Va.*

1—Thirty years. 2—No. 3—No. 4—5,000. Remarks : Do not consider it possible for transitory mania to exist as an idiopathic disease. Yet I would not deny the possibility of transient maniacal phenomena as a consequence of temporary physical conditions.—*Pliney Earle, M.D., Supt. of the State Lunatic Hospital at Northampton, Mass.*

1—Twelve years. 2—No. 4—About 1,500. Remarks : Cannot say what is or is not possible.—*H. P. Stevens, M.D., Retreat for the Insane, Hartford, Conn.*

1—Forty-two years. 4—8,000.—*H. A. Buttolph, M.D., Morris Plins, New Jersey.*

Accompanying this reply is a highly interesting letter, in which the author doubts the theory of transitory mania, and thinks the term should be dropped and "insane impulse" be used in its place.

1—Twenty-eight years. 2—Have never seen such a case. 3—No. 4—3,000. Remarks : Neither experience nor reading lead me to think it possible. The only cases which have come to my knowledge claimed as transitory mania have been supported by very questionable evidence. —*J. P. Bancroft, M.D., Supt. Insane Asylum, Concord, N. H.*

The gentlemen who have kindly furnished me with the foregoing statistics are highly esteemed by the medical profession for their learning in psychological medicine, and their statements are entitled to the fullest confidence as authority upon the subject; and, in justice to offended law, we must conclude that the theory of *mania transitoria* to the criminal is like the signal of the mariner, far out to sea upon a sinking vessel, with no reasonable help in view; it is the only hope of relief, when human sympathy alone comes to the rescue, powerless to save the innocent from the perils of the wave, yet holds the guilty in the embrace of social life, though he can never be fully restored to the confidence of the people.

In the revised edition of the "Medico-Legal Papers," page 189, Dr. Wm. A. Hammond says : "The sympathetic system of nerves has a most important office to perform in the organism, and one which in its relations to the subject is of very great moment. It is the organ by which the size of the blood-vessels is regarded." And on page 190, xxi : "Now, what is the condition known as transitory mania? 1. It may be defined as a form of insanity, in which the individual, with or without the exhibition of previous *notable* symptoms, and with or without obvious exciting cause, suddenly loses the control of his will, during which period of non-control he commonly perpetrates a criminal act, and then as suddenly recovers, more or less completely, his power of volition. 2. Attentive examination will always reveal the existence of symptoms precursory to the outbreak which constitutes the culminating act, though they may be so slight as to escape superficial examination. (*a*) The hypothesis, therefore, that a person may be *perfectly sane one moment, insane the next*, and then again *perfectly sane* in a moment, is contrary to all the experience of psychological medicine."

Page 185 : "An essential feature of the definition of insanity is that it depends directly upon a diseased condition of the brain."

"Medico-Legal papers," page 221 : "Whatever may be said by the pure psychological school of philosophers, the world is indebted to physicians and physiologists for the only true philosophy of mind, namely, that instead of being a simple entity, an independent source of power and self-sufficient cause of causes, it is dependent on a material organ for all its manifestations."

"Mental power is but an organized result, matured by insensible degrees in the course of life, and as much dependent on the nervous structure as the function of the liver is on the hepatic structure."

Report of Abner Rogers, Jr., indicted for the murder of Charles Lincoln, Jr., tried before the Supreme Court of Massachusetts, page 104.

"And first as to the point of the State's attorney, that the prisoner's offense, if not amounting to murder, may yet constitute manslaughter. That as manslaughter is murder upon provocation, or under sudden excitement, so murder *on insane* or *partially controllable impulse*, may be no more than manslaughter." As there can be no such thing as "partially controllable impulse," no great wisdom is presented in this theory, nor evidence of careful investigation of mental disease, nor does it enlighten the jury.

A great statesman of the old country, Mr. Burke, said that "the soul of government lies in the jury-box." But jurors should be well-informed, and competent to pass upon and determine between questions which relate to facts and those which relate to law ; and experts upon questions of insanity should be able to make such clear statements before them that there should be no difficulty in understanding the true import of their meaning. But no one can prove to any jury that *impulse*, or *emotion*, is a disease which only finds a remedy in some criminal act, as homicide, and then is instantly and forever cured.

As a matter of physiological interest connected with this subject, and showing some of the functions of the brain, Luys (page 53) remarks : "Subject, who had been long deprived of an upper limb, in the case of disarticulation of the shoulder, there existed in certain long disused regions of the brain, coincident, very distinctly localized atrophies. I have, moreover, demonstrated that the atrophied regions of the brain are not the same in the case of the amputation of the leg as in that of amputation of the upper limbs."

The researches of Feitsch, Hitzig, Panier, Brown-Sequard, Bartholow, and many others, have shown that by applying electric excitement in the region of the gray cortex motor, reaction in isolated groups of muscles are determined ; that at will we may cause the eyes, tongue and neck to move.

The period of incubation of reflex nervous action varies, and the inhibitory force of nervous activity is the only important factor which will or can determine the duration of physical disturbance as it relates to any specific action dependent upon nervous energy. This is often witnessed in cerebral apoplexy, or injuries to the brain from extraneous causes, and shows in the simplest manner the correlation between cause and effect. And when we apply these truths to the inhibition of mental action, we shall be able to prove by physiological facts that spontaneous development of disease is impossible, as it relates to growth and decay at the same moment.

In a work entitled "Plain Talk about Insanity," by T. W. Fisher, M.D., page 86, we find the following, viz: "Epileptics are known to be subject to attacks of frenzy. This knowledge makes physicians careful, in cases of unexplained violence, to search for some trace of epilepsy, vertigo, or *petit mal*, in the previous history of the suspected person, and it is often found." He states that Dr. Krafft Ebbing "distinguishes seven different groups of conditions, under any of which transitory mania may occur," viz :

1. The state of dreaming.
2. Different kinds of intoxications.
3. Delirium of febrile maladies.
4. Transformation of neuroses.
5. Transitory psychoses.
6. Pathological passion.
7. Transitory intellectual troubles at child-birth.

Alas! *parturiunt mentes*, nothing having been brought forth by his labors but confusion.

This classification and subdivision of the etiology of transitory mania is too vague to secure importance with the careful reader ; and if it shows anything, it proves the error of the author's theory and the incorrectness of his conclusions, for a manifestation of frenzy or great mental excitement connected with either of these physical conditions can only be rationally considered as symptoms of some existing malady.

If these conditions collectively have any central meaning, they refer to epilepsy, not to mania transitoria as an idiopathic disease, and the words of the author himself cannot lead one astray from this conclusion, for he states: "All these conditions of transitory disorder may prove very difficult to estimate, because the direct examination of the accused only affords negative results."

Clouston, on "Mental Diseases," page 162, remarks: "I think cases of mania transitoria result from the following causes: Most of them are *epileptiform*, are, in fact, of the nature of mental *epilepsy*. All the symptoms may be seen in the incubation of febrile and inflammatory complaints, such as scarlet fever, typhus and typhoid, local inflammations, etc."

He speaks of visiting a person who was very wakeful, and was laboring under some peculiar mental aberration that came on suddenly, and says, since then, when he has similar cases, he asks himself, "Is it a case of mania transitoria?" and then states he has seen many similar cases in asylums, *especially among epileptics.*

Edward C. Mann, in his recent work on Psychological Medicine, page 122, says: "There are certain cases familiar to all specialists in insanity, which suffer from impulsive insanity, with a homicidal or suicidal monomania. These patients, without appreciable disorder of the intellect, are impelled by a terrible *vis a tergo*, a morbid, *uncontrollable* impulse to desperate acts of suicide or homicide." He also speaks of a patient under his care "who would voluntarily enter an asylum and remain there until the morbid impulse had passed away."

While it should be remembered that an isolated case is not

sufficient evidence to prove an important fact, the case cited shows that the proper exercise of the will was sufficient to control what the author says is an *uncontrollable impulse*, for the patient went away;of his own accord and escaped the reality of his dreadful forebodings.

Again, on page 126, he says, as a climax to the discussion of *inebriety:* "In these cases also the mental disorder is of a sudden and *transitory* character, not preceded by any symptoms calculated to excite suspicion of insanity." "It is a *transitory mania*, or *sudden paroxysm*, without antecedent manifestation," the duration of the morbid state being short, and the cessation sudden. Such attacks are transient in proportion to their violence, and transition occurs on the completion of the act of violence." "Clearly allied to this state of which I have been speaking is that peculiar psychological state, the trance state, which also occurs in inebriety."

Trance is an exceedingly rare condition—so rare, indeed, that ten thousand physicians may enjoy a large practice for many years, and not one of them ever witness a typical case.

Alexander Bain, in his work upon the "Intellect" has many valuable thoughts connected with this subject, which are in accordance with the views I have expressed.

When carefully considered, this will be found only to relate to some manifestation or symptom of a disease known to exist, and has no bearing upon the question of transitory mania as an independent or idiopathic disease. Transitory mania as the result of inebriety has nothing in common with trance; trance being unquestionably an idiopathic condition in which the patient remains as if Leothe had breathed over the entire organism a feeling of repose, and left the mind oblivious to its own consciousness to revel in dreams and ecstacies. Here the author is clearly in error, for when he speaks of trance being closely allied to transitory mania in connection with inebriety, he incorrectly classes trance with one of the many phases of epilepsy.

Clouston, in his work on Mental Diseases, discusses the subject of "impulse" in a very fair tone of argument, and favors the theory that it is "irresistible," but says nothing of "transitory mania," and leaves the correctness of his syllogistic reasoning to the same criticism of analysis I have given to Mann; for the latter has not proven that transitory mania is an idiopathic disease, and the former has not shown that "impulse" is uncontrollable.

The theory is easy to accept, but in itself proves nothing. The refractory horse is impulsive, but by careful training becomes submissive; by kind treatment the wild animal obeys the voice of its keeper. All impulses are the result of cerebral excitement, which by the control of psychological forces may be overcome in man as in the animal, and the person who does not attempt to control himself commits a great moral wrong, and the "sin of omission" rests upon him.

"'Tis education forms the common mind," and even where impulse is an "inheritance," the force of moral training and individual discipline can and should hold it in abeyance as far as it relates to criminal action. The mother's love controls her sullen and sulky child until his eye is full of laughter and his cheek dimpled with smiles. Psychopathy controls the turbulent forces of his inhibitory nature, and he is subdued. Man is but a child of "larger growth."

In this connection we may ask what are the bearings of the developmental theory, and the theory of evolution, upon the ethics, morals, health, and law of the present age? A man is either amenable to himself or to some established law, with precepts his guide, but law his rule, and if healthful laws are not administered, crime is either evolved or developed for evolution, and the highest interests of society are sacrificed to caprice and misguided judgment as an unavoidable sequence. It is only through the reasoning by induction, inference and comparison with concrete things and demonstrative principles, we are in anywise able to judge of the subtle, controlling forces of mentality, as witnessed in the agency and effect of electricity, produced by induction or chemical reaction, electrolysis, Farradism, magnetism, etc. By the force of constructive genius applied to mechanical arts, we are able by a simple touch to illuminate cities or sound the alarm of approaching danger. But as this paper is not intended to enter into an exhaustive discussion of this subject, connected with mind, I will leave this train of reasoning to be matured by others.

Mann (page 43) says : " In epilepsy the most internal part of the ascending parietal convolution of the brain has been found to be atrophied and indurated to cartilaginous consistence as far as its embrochure in the fissure of Sylvius." If we state as a pathological fact that it is a disease without a pathology, we shall fail to keep in the path of scientific truth and medical learning ; and the more we investigate this question, the more certain will become our conviction that the plea of transitory mania is a mere subterfuge to enable the criminal to escape legal responsibility.

" Epileptic vertigo, which a person may have had for years without suspicion of its true nature, on the part of himself or his friends, is very fruitful of mental disturbances. The *irritation*, we may call it, may at any time seize the higher centres of the brain, instead of the lower, producing delirium as transient as the vertigo. In this transitory mania—for it is such—an act of violence may be done for which the patient is utterly irresponsible."—*Fisher*, page 27.

This author keeps in the same trodden path of all other writ-. ers upon this subject, who never fail to associate transitory mania with epilepsy, where it must for ever rest as a manifestation of disease rather than a disease entitled to independent nosonomy. Motive and volition cannot be separated in any conscious act performed, for there can be no motive without the exercise of the will; and if it can be ascertained that a motive existed for the commission of a crime prior to its committal, then the act must carry with it the responsibility of the actor.

All the leading authorities upon medical jurisprudence and mental diseases, as Maudsley, Ray, Esquirol, Hamilton, Spitzka, Bucknell, Tuke, Beck, Taylor, Browne, Mann, and Luys upon the "Functions of the Brain," agree upon the legal proposition that any party competent to distinguish right from wrong stands in the presence of the law as sane, and is responsible for his acts.

Luys (page 421) says : "The corticle periphery surrounding the optic thalami becomes intellectualized in some way to serve as exciting material for the activity of the cells of the corticle substance. These are the open gates by which all stimuli from without destined to serve as *pabulum vitæ* for these same corticle cells pass, and the only means of communication by which the regions of psychical activity come into contact with the external world."

From this view of the author, there appears to be no doubt of a cerebral localization of psychic and intellectual activity, and " that the sensory organs have a receptive organ in some way adapted to it in the central regions."

I have stated I had no objection to the theory of transitory mania as a symptom of some existing form of insanity, or as a manifestation of epilepsy; but it would be subject to less objection to consider this erratic condition of mind a *nomadic spasm* of localized forces of the brain; and if it cannot be shown that some lesion of the brain existed prior to the outburst of criminal excitement, then the party must be held responsible for any act of violence he may have committed; for the act would be voluntary, and all voluntary acts can be resisted.

When this view of the question of transitory mania shall become established law, farcical court dramas, where the insane act is played, will no longer be brought before the public, with all the mockery and satire upon justice and rhapsody of polemic pleading in such cases.

Public morals and common decency demand the restitution of common sense of parties interested in court proceedings in all insane criminal actions, that juries may be enlightened, and not confused by too much technical law and uncertain testimony.

If a person commits murder while intoxicated, the law does not hold him guiltless, because he voluntarily placed himself in that condition which rendered him unconscious of the act he was committing. So if one commits murder while laboring under any great impulsive excitement, he becomes and should be held by the law as responsible; for all such persons can control their feelings, and would do so if there was a certainty of punishment before them from which they could not escape. And here is where we see the psychopathic influence as a restraining force in the prevention of crime; for any specific law placed upon the statutes affecting crime, which makes its punishment sure, would become a psychopathic force to restrain criminal action, and more attention should be given to this subject in connection with the study of forensic medicine.

Gen. Grant, in a conversation with Prince Von Otto Bismarck, alluding to the attempted assassination of King William, said: " Although at home there is a strong sentiment against the death penalty in cases of spasmodic insanity, and it is a sentiment which one naturally respects, I am not sure but it should be made more severe, rather than less severe. Something is due to the offended as well as the offender." "That," replied the Prince, "is entirely my view. My convictions are so strong that I resigned the government of Alsace because I was required to commute sentences of a capital nature." Let there be a few prompt examples of laws faithfully executed, and the psychopathic effect will be so great that homicidal lawlessness of *cranks* will soon cease.

As objects multiply upon which we reason as we advance in intellectual development, *pari passu*, so a series of new forces spring up, which become factors in determining processes of judgment; hence we find equally well-educated men to differ in their opinions upon questions involved in controversies, and what seems material and pertinent to one is objective to another in its logical sequence and application.

DEFINITIONS OF INSANITY.

The legal definition of insanity is, that it is "a condition of mind which renders any party incapable of judging between right and wrong in any particular act at the time it is committed." A philosophical definition declares, "Insanity is a mental state in which acts of conception, judgment, or reasoning, persistently express themselves as different from the states of feeling and modes of thought usual to the individual in health" (Combe).

Such conditions as these render the patient legally an irresponsible being, and unfit him eventually for the performance of the social and political duties of life, for behind the act remain evidences of disordered intellect. This conclusion is in accordance with the expressed views of Esquirol, Tuke, Winslow and Bucknill.

A pathological definition of insanity is, an aberration and erratic condition of mind, dependent upon a morbid state of the cerebral nerve centres which control mental activities; and it

varies in form as the location of nerve centres varies that are
affected, and the changes produced in mental manifestation by
pressure, over-stimulation, or lack of nourishment of the brain,
are often analogous to the different forms of paralysis and aphasia,
according to the localized disturbances. Under such a condition
molecular forces are interrupted, and mental activity, as the intellectual factor is left struggling through abnormal media to
give expression of the disturbance to the outward senses.

["The phenomena of moral responsibility, considered as a
purely physiological synthesis of all nervous activities, consists
in a series of regular processes, executed by the organism at its
own expense, and resulting from the harmonic consensus of all
its parts. Moral sensibility finds also in the intervention of intellectual activity a new power which excites it, makes it active,
and maintains it in a perpetual state of erethism."]—*Luys, page
109*.

While this parenthesis refers to the social conditions of life, the
truths it contains stand at the gateway that leads up to the fundamental truths of the conditions which impose legal responsibility. This also accords with the modern researches of those
who have given much careful attention to neurology and the
functions of the brain, and is the echo of the voice of science,
reverberating through all the avenues of thought and the intri-
cacies of life, arousing the force of mental activity, the *vividu vis
animi*, which awaken the consciousness of the errors of judgment manifested in crime.

Ray, in his "Jurisprudence of Insanity," states that " the pro-
pensities and sentiments are also integral portions of our mental
constitution; and enlightened physiologists cannot doubt that
their manifestations are dependent upon the cerebral organiza-
tion."

As far back as the days of Paracelsus and Democritus, Hypocrates expresses his views of cerebral physiology and of the pathology of insanity as follows, viz : " Men ought to know that from nothing else but thence [referring to the brain] come joys, despondency and lamentations."

Says Aitken : "To consider the subjective phenomena which collectively, in their various manifestations, constitute mind, an immaterial essence, as liable to disease apart from all derangement of the material organ, the instrument with which it is so closely and indissolubly united, is to believe in a most incongruous, unphilosophical, unphysiological doctrine. The more consistent theory is that which is known as the cerebral theory, now entertained by most of the eminent physicians who have made insanity a special study."

Dr. Boyd records the singular fact that almost invariably the weight of the left cerebral hemisphere exceeds that of the right by at least an eighth of an ounce.

The weight of the brain under different forms of insanity has been found as follows, viz : In mania the brain weighs 54 oz. 11½ drams ; in monomania, 51 oz. 11⅜ drams ; in dementia, 50 oz. 5 ⅔ drams ; in general paralysis, 49 oz. 12 ⅘ drams. This tabular statement shows most conclusively that the true pathology of insanity is found in the encephalon. The specific gravity also varies in the white and gray matter.

Aitken alludes to *irresistible* impulse, but thinks *unresisted* a better term.

I introduce these facts to show the difference between insanity as a fact and the theory of insanity in *mania transitoria*, the latter having no pathology; and while the logic in any particular case may be correct, the premises, *a priori*, being false upon which to base a plea, the conclusions must necessarily be wrong and verdicts will be rendered in favor of the criminal.

Clouston (page 232) says: " Professor Benedick, of Vienna,

showed at the International Medical Congress of 1881, in London, a number of brains of habitual criminals, who, he affirmed, had their convolutions arranged in a certain simple form peculiar to the criminal classes, so that on seeing such a brain he could tell the ethical tendencies of the person to whom it belonged, just as one can tell a dog to be a bull-dog by his jaws."

When one is suffering from dementia, the listless, incoherent condition is interrupted by a sudden loud tone of voice, that causes a vibration of certain cerebral nerve centres, which is exhausted in a vocal response, while the party is wholly unconscious of its meaning or of any of the surroundings.

Bucknill and Tuke, in their work upon "Physiological Medicine," page 443, referring to the pathology of insanity to cerebral disturbance, state the following facts : "Greeding, in 216 cases, found the skull unusually thick in 167, the *dura mater* adherent to the cranium in 107 cases, the *pia mater* thickened and opaque in 86 out of 100 cases of mania, and beset with hydatids and spongy bodies in 92 out of 100 cases. The choroid plexus was found healthy in these respects in only 16 cases out of 219. Merkel noticed the increased density of the cerebral substance" in insanity.

Sœmmering and Arnold confirm these observations, and Pascal declares that all mental diseases are the effect of "morbid alterations in the brain and spinal cord." I will also call attention to the views of Virchow, as expressed in his " Cellular Pathology," in support of the same conditions. While his views are not given in connection with any discussion of this subject directly, they have an important bearing upon the theory that all insanity is dependent upon cerebral changes, which may be slow and progressive in nature, as when melanoid tumors or gliomata press upon the corpus striatum, "that have their origin in the neuroglia of the interstitial connective tissue," and many other forms of change.

It is not logical to say because the song of the bird is not found in its throat, *ergo*, the anatomical structure of the throat has nothing to do with the musical notes; nor would it be reasonable to say because specific lesions were not found, *ergo*, the anatomical changes of the eucephalon had nothing to do with mental diseases.

Most of the late writers agree to this proposition, and daily experience proves its correctness: " We must be careful not to underrate their importance on account of the occasional absence of anatomical changes after death, and to conclude that for this reason such anatomical lesions, when present, may not be the cause of the mental disorder."—*Griesinger, page 291.*

While man does not possess a prescient mind that would enable him to unmask the speculative thought which underlies and controls much of human action, a knowledge of the force of mental activity must be gained through a comparison of the negative with the active life manifested in physical efforts. The General upon the battle-field gives the command, and the force of the law of obedience brings contending armies together in deadly strife, and human life is sacrificed. This is a fair illustration of the force of the law of psychopathy, and will enable us to judge more fairly of the contingent results of daily action, and show how the positive execution of law will tend to lessen criminal action.

The clinical importance of the subject of psychology and biology, as questions of science, seems to have been overlooked by our teachers of medicine, and as yet these branches have not been assigned a place in the college curriculum; and whatever of public attention may have been directed to this department of learning, none but the flippant writers upon the stage of experiments have undertaken the explanation of the curious effects mind of upon mind, until quite recently, notwithstanding Baron

Reichenbach's experiments in odic forces and biology nearly thirty years ago.

> "I know
> That where the spade is deepest driven,
> The best fruits grow,"

and make the following quotations from the work of Griesinger upon " Mental Pathology :"

" Emotion, when transitory, and occurring in previously healthy organisms, is speedily calmed ; when, however, bodily disease is already present, and when the causes are long-continued, there generally arise many complicated disorders of the organic mechanism, which the simple cessation of the emotion cannot as quickly terminate."—Page 40.

" Very much depends upon the duration and intensity of the phenomena, whether we consider the mental state as morbid."—Page 44.

" We frequently see in subjects who, up to that moment, have been in the actual or at least apparent enjoyment of perfect health, just as in some of those cases in which there is developed a suicidal tendency, attacks of most violent anxiety with obscuring of consciousness suddenly show themselves, accompanied with frightful hallucinations, during which the patient, in the blindness of his fury, seeks to slay all who come in his way. These cases, which, judged by their symptoms, appertain more, it is true, to *mania*, but which in their psychological relations represent violent fits of melancholic anxiety, and especially morbid negative emotions, possess, in their want of any moral cause, a great analogy to those sudden fits of profound anxiety and severe mental suffering which have sometimes been witnessed as precursors of epileptic attacks."—Page 184.

Under this head come " those cases in which those homicidal impulses suddenly, and without external motive, arise in persons who have been hitherto of a lively, joyous, and loving disposition, and incessantly intrude themselves upon their thoughts."—Page 185.

Here the author refers to several notable examples of homicidal impulse: "A distinguished chemist, tormented with a homicidal impulse, would often prostrate himself before the altar and implore the Deity to deliver him from the atrocious propensity," etc. Also refers to Catherine Olhaver, to a nurse, and the wife of a shoemaker, who were "seized with an *almost* irresistible impulse."—Page 187.

He also speaks of the "habitual perversion of the feelings with impulsive fits of anger, without any derangement of the intellect," and cites the case of "an only son, brought up under the eyes of a weak and indulgent mother, who early acquired the habit of *yielding* to all his caprices and to all the impulses of a restless and ardent temperament" (page 190), and on page 208, example xxxv., refers to "paroxysms of fury," and as an illustration relates the story of the Swabian peasant, with which every student of psychological medicine is familiar ; but in none of these cases has he shown that the impulse is *irresistible*.

Upon such cases as these is based the testimony in favor of the theory of ephemeral "transitory mania," with which our court calendars are filled ; and these cases even do not prove either the conscious or unconscious irresistible nature of the malady; nor does this classical writer claim or attempt to prove that transitory mania exists as an idiopathic disease, and common intelligence forbids such labor, as an effort against justice and moral rectitude.

"The fundamental disorder in mania, the irritation upon the motory side of the soul-life, exhibits itself, first of all, in this sphere, as a high degree of mental excitement, with restlessness, impetuous and violent desires and actions." "The pleasure in

loud speaking, in shedding blood, etc., may show itself in those
violent and boisterous ways, and these results, fixed or *transitory*
conditions which, according to the predominance of this or that
desire, are known under the name of kleptomania, homicidal
mania, etc."—Page 197.

" Generally, from the commencement of insanity, or at least
very soon, the quantitative increase and exaltation of thought are
so great that there results a restless and constant succession of
isolated ideas which have no intimate relations with each other,
and constantly change their combinations, are very *transitory*, or
of a very fragmentary nature."—Page 199.

" Whether, and to what extent, certain directions of the will
and impulses in the insane, particularly such as lead to criminal
acts, are irresistible, is a question which can scarcely be answered
with certainty."—Page 55.

" As to the *invasion and course* of mania, it is observed sometimes as a pure and independent form of mental disease, as a
stage of development in the successive series of mental disorders ; sometimes *transient* attacks of mania, or more correctly of
fury, occur in individuals who are already subjects of profound
mental disease." " In *epileptics*, also, it is not uncommon to ob-
serve attacks of mania which are often characterized by a high
degree of blind jury and ferocity. Sometimes they immediately
follow the *epileptic* attacks."—Page 203.

The same field of research has been traversed by this learned
and classical writer that has so repeatedly been investigated by
others, without any apparent effort to prove how such a condition
can occur possibly under the laws of psychopathy or psychosis,
as understood at the present time; and until some more positive
evidence of its existence shall be presented, the duty I owe to
society will make it my aim and urge me to oppose the theory
of *transitory mania* as an ogre too mythical to be brought be-

fore courts and juries. The baneful effects of this theory upon public morals are matters of record and familiar to the judiciary of our country.

In discussing this subject I have endeavored to avoid sacrificing facts to either ethics or rhetoric, holding that the cognate sciences are the factors to be relied upon to explain by comparison and illustrate the phenomena of cause and effect; and have endeavored to show that while the remote causes of insanity multiply with daily experience, pathologists have established the fact that the disease itself is dependent upon cerebral lesions and disturbances of nerve centres of the encephalon, which may be either primary or dependent upon reflex action, as in hysteria, dyspepsia, disease of the heart, spine, liver, and kidneys. "The anatomical changes which indicate insanity—that is, which produce psychical anomalies during life—are naturally sought for within the cranium, in the brain and its membranes."—*Griesinger*, page 290.

Thus, with the preponderance of evidence in our favor, we are forced to conclude that it is impossible for transitory mania to hold a place in medicine, *per se*, and that the theory is surrounded by more empiricism of law and medicine than any other humbug of modern times.

In conclusion, whatever may be thought of our method of life and our efforts in the various walks of a chosen profession, may we never forget to cherish a love for justice, reverence for law, respect for the rights of others and sympathy for the unfortunate in mind; and when our labors shall cease, may we have a consciousness of having performed our duties faithfully and well, without fear or prejudice towards any.

> " Whatever creed be taught or land be trod,
> Man's conscience is the oracle of God."

www.ingramcontent.com/pod-product-compliance
Lightning Source LLC
Chambersburg PA
CBHW021640270326
41931CB00008B/1097